SCHOLASTIC

W9-ANY-644

Success With

Math

New York • Toronto • London • Auckland • Sydney
Mexico City • New Delhi • Hong Kong • Buenos Aires

Teaching *Resources*

State Standards Correlations

To find out how this book helps you meet your state's standards, log on to **www.scholastic.com/ssw**

Cover design by Ka-Yeon Kim-Li
Interior design by Ellen Matlach Hassell
for Boultinghouse & Boultinghouse, Inc.

ISBN 978-0-545-20071-4

29   40   22 21 20

Contents

About the Book

"Nothing succeeds like success."
—Alexandre Dumas the Elder, 1854

And no other math resource helps kids succeed like *Scholastic Success With Math*! For classroom or at-home use, this exciting series for kids in grades 1 through 5 provides invaluable reinforcement and practice for math skills such as:

❏ number sense and concepts
❏ reasoning and logic
❏ basic operations and computations
❏ story problems and equations
❏ time, money, and measurement
❏ fractions, decimals, and percentages
❏ geometry and basic shapes
❏ graphs, charts, tables ... and more!

This 64-page book contains loads of challenging puzzles, inviting games, picture problems, and clever practice pages to keep kids delighted and excited as they strengthen their basic math skills.

What makes *Scholastic Success With Math* so solid?

Each practice page in the series reinforces a specific, age-appropriate skill as outlined in standardized tests. These skills help kids succeed in daily math work and on standardized achievement tests. And the handy Instant Skills Index at the back of the book helps you zero in on the skills your kids need most!

Color the Basket

Name _____ Date _____

Count the number of dots or triangles in each shape. Then use the Color Key to tell you what color to make each shape. (For example, a shape with 7 dots will be colored green.)

Extra: On the back of this sheet of paper, draw a basket filled with six things you would carry in it.

Color Key
6 = yellow
7 = green
8 = brown
9 = red
10 = green

Number User

Name _____ Date _____

I use numbers to tell about myself.

1. _____
   MY STREET NUMBER

2. _____
   MY ZIP CODE

3. _____
   MY TELEPHONE NUMBER

4. _____
   MY BIRTHDAY

5. _____
   MY AGE

6. _____
   MY HEIGHT AND WEIGHT

7. _____
   NUMBER OF PEOPLE IN MY FAMILY

I CAN COUNT UP TO

8. _____

Frog School

Name _____ Date _____

At Frog School, Croaker Frog and his friends sit on lily pads.
Are there enough lily pads for all the frogs in Croaker's class?
Yes ____   No ____
Draw lines to match the frogs with the lily pads.

Extra

How many frogs need lily pads? _____.

Odd and Even Patterns

Name _____ Date _____

A pattern can have two things repeating. This is called an "AB" pattern.

1. Look around the classroom. What "AB" patterns do you see?
   Draw one "AB" pattern in the box.

2. Use red and blue crayons to color the numbers in the chart using an
   "AB" pattern.

Hundred's Chart

1	2	3	4	5	6	7	8	9	10
11	12	13	14	15	16	17	18	19	20
21	22	23	24	25	26	27	28	29	30
31	32	33	34	35	36	37	38	39	40
41	42	43	44	45	46	47	48	49	50
51	52	53	54	55	56	57	58	59	60
61	62	63	64	65	66	67	68	69	70
71	72	73	74	75	76	77	78	79	80
81	82	83	84	85	86	87	88	89	90
91	92	93	94	95	96	97	98	99	100

Use this rule:
1 = red
2 = blue
3 = red
4 = blue, and so on

The blue numbers are **even numbers.** They can be split evenly into 2 whole numbers.

The red numbers are **odd numbers.** They cannot be split evenly into 2 whole numbers.

Classroom Garage Sale

Name _____ Date _____

Tolu's class did some spring cleaning. Then they had a garage sale. They sorted the things they were selling. Sort these objects into like groups. Draw the items of each group on one of the tables below.

_____   _____   _____

Below each table, write a label for the group.

Snowflakes on Mittens

Name _____ Date _____

Estimate how many snowflakes are on each mitten.

For the first mitten, skip count by 2s to find out.

(You can circle groups of 2.)

For the second mitten, skip count by 5s to check your answer.

(You can circle groups of 5.)

Extra

Would snowflakes really wait for you to count?

Explain your answer:

Sorting Treats

Name _____ Date _____

Look at the Halloween treats below. Cut apart the boxes. Then sort them into 2 piles. One pile is for numbers greater than 10. The other pile is for numbers less than 10.

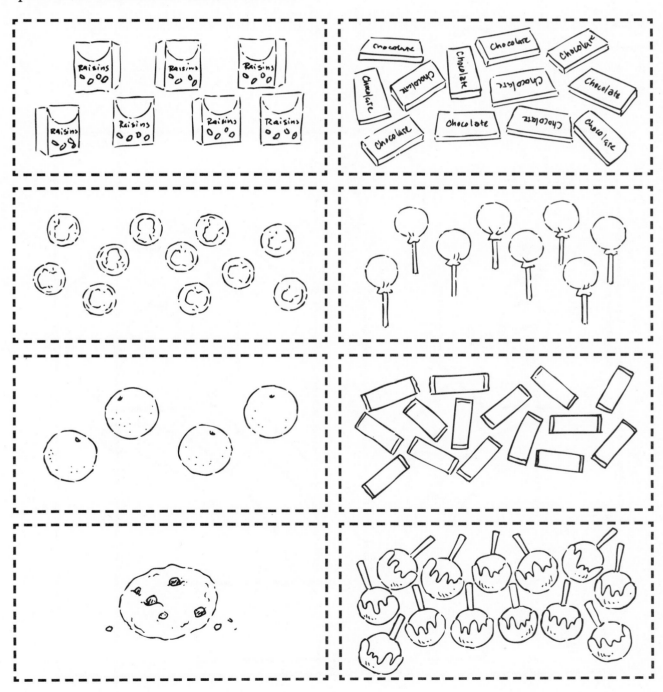

Flowers in a Pot

Name _____ Date _____

Count the dots in the boxes. Then color the matching number word.

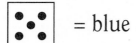 = green   = yellow

= red   = purple

Extra: Use bright colors to draw a pot of flowers on the back of this sheet of paper.

= blue

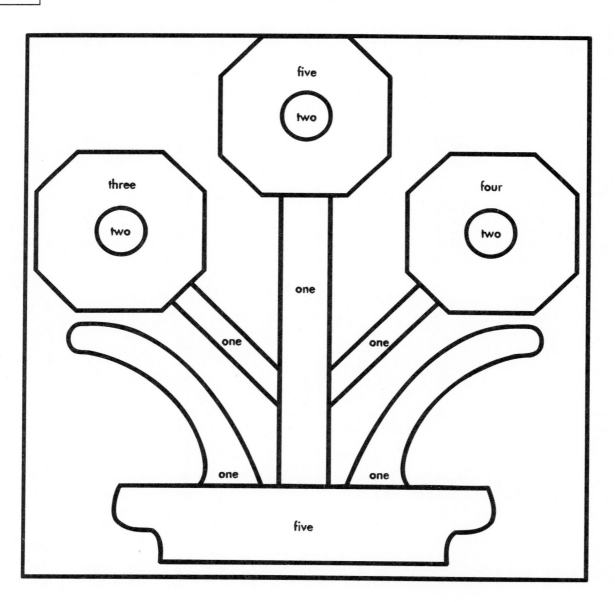

Sign Shape

Name _____ Date _____

Street signs come in different shapes. Use string to form the shapes below. Work with a partner. Answer the questions below about the shapes, too.

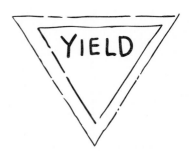

What shape is this sign? _____

How many sides does it have? _____

What shape is this sign? _____

How many sides does it have? _____

What shape is this sign? _____

How many sides does it have? _____

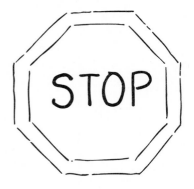

What shape is this sign? _____

How many sides does it have? _____

Bird Feeder Geometry

Name _____ Date _____

It's spring! The birds are coming back. Kwaku and his mother made
two bird feeders. What shapes can you find on their feeders? Write
your ideas on the lines. _____

_____

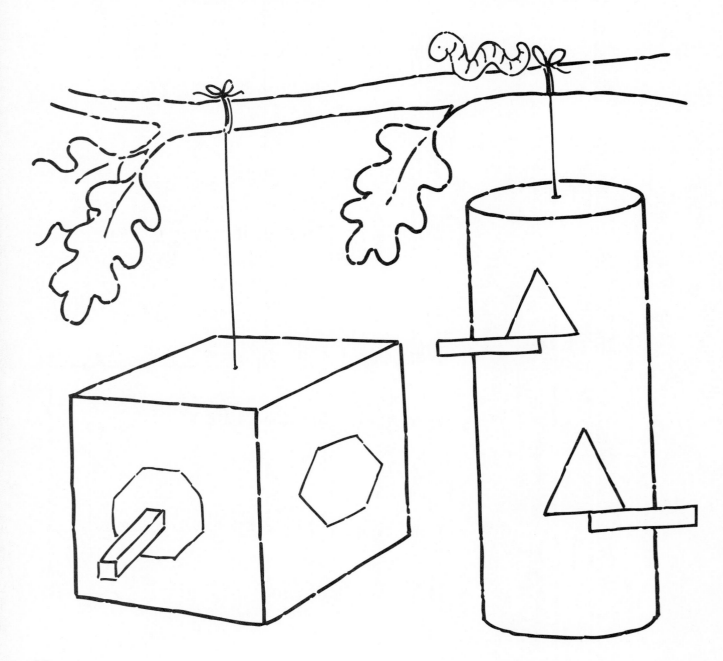

Shape Study

Name _____ Date _____

"Symmetry" exists when the two halves of something are mirror images of each other. Look at the pictures below. Color those that show symmetry. (Hint: Imagine the pictures are folded on the dotted lines.)

Complete the drawings below. Connect the dots to show the other half. (Hint: The pictures are symmetrical!)

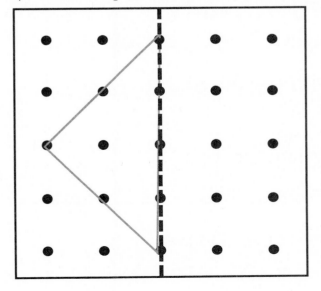 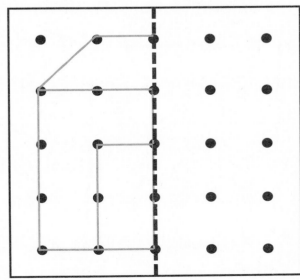

15

Picking Out Patterns

Name _____ Date _____

On the 100th day of school, everyone in Pat's class picked out patterns on the 100 Chart. Look at the chart below.

1	2	3	4	5	6	7	8	9	10
11	12	13	14	15	16	17	18	19	20
21	22	23	24	25	26	27	28	29	30
31	32	33	34	35	36	37	38	39	40
41	42	43	44	45	46	47	48	49	50
51	52	53	54	55	56	57	58	59	60
61	62	63	64	65	66	67	68	69	70
71	72	73	74	75	76	77	78	79	80
81	82	83	84	85	86	87	88	89	90
91	92	93	94	95	96	97	98	99	100

Find and finish the pattern starting with 2, 12, 22

_____

Find and finish the pattern starting with 100, 90, 80

_____

Find and finish the pattern starting with 97, 87, 77

_____

Find and finish the pattern starting with 11, 22, 33

_____

Fall Leaf Patterns

Name _____ Date _____

Fred made a pattern using the leaves below. You can make patterns, too. First, color the leaves as shown in the small boxes. Then color and cut out the bigger leaves. Use the leaves to make patterns.

Red	**Orange**	**Yellow**
maple leaf	oak leaf	elm leaf

Valentine Symmetry

Name _____ Date _____

Janis folded all her valentines in half. Some were symmetrical—one half matched the other half. Some valentines were not symmetrical.

Cut out the valentine shapes below. Fold them in half by making a crease that runs from top to bottom. Which ones are symmetrical?

_____   _____

_____   _____

Fold the symmetrical shapes in half so that you make a crease running from left to right.

Which shape is symmetrical this way? _____

18

Patterns of Five

Name _____ Date _____

Look at the number chart below. Starting with 1, count 5 squares. Color in the fifth square. Then count 5 more squares and color in the fifth square. Keep going until you reach 100.

Hundred's Chart

1	2	3	4	5	6	7	8	9	10
11	12	13	14	15	16	17	18	19	20
21	22	23	24	25	26	27	28	29	30
31	32	33	34	35	36	37	38	39	40
41	42	43	44	45	46	47	48	49	50
51	52	53	54	55	56	57	58	59	60
61	62	63	64	65	66	67	68	69	70
71	72	73	74	75	76	77	78	79	80
81	82	83	84	85	86	87	88	89	90
91	92	93	94	95	96	97	98	99	100

Tally marks can be arranged in groups of five, like this: ⵑⵑⵑ ⵑⵑⵑ ⵑⵑⵑ Then you can count by fives.

Count how many girls and boys are in your class. Draw tally marks in groups of five.

Girls: _____ Boys: _____

Now count the total number. Write the totals here:

Girls: _____ Boys: _____

19

Ladybug Dots

Name _____   Date _____

Every year, ladybugs hibernate when the weather gets cool. Count the dots on each ladybug wing. Then write an equation to show the total number of dots each ladybug has. The first one has been done for you.

3 + **3** = **6**

_____ + _____ = _____

_____ + _____ = _____

_____ + _____ = _____

_____ + _____ = _____

_____ + _____ = _____

Extra Write the sums in order, from lowest to highest.

_____ _____ _____ _____ _____

What pattern do you see?

_____

Pattern Block Design

Name _____ Date _____

How many total pieces are in this pattern block design?

2 + 2 + 1= _____

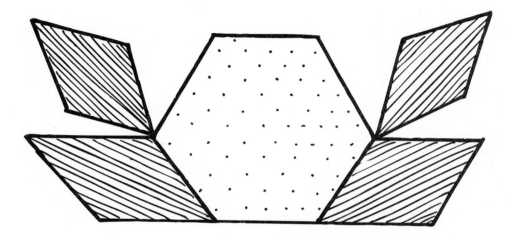

Now make your own design by drawing 5 pattern blocks. Connect the blocks to form a pattern different from the one above. You may want to use a block pattern more than once.

Write an equation to show how many of each shape you used.

Equation: _____

Animal Mystery

Name _____ Date _____

What kind of animal always carries a trunk?

_____

To find out, solve the addition problems. If the answer is greater than 9, color the shape yellow. If the answer is less than 10, color the shape gray.

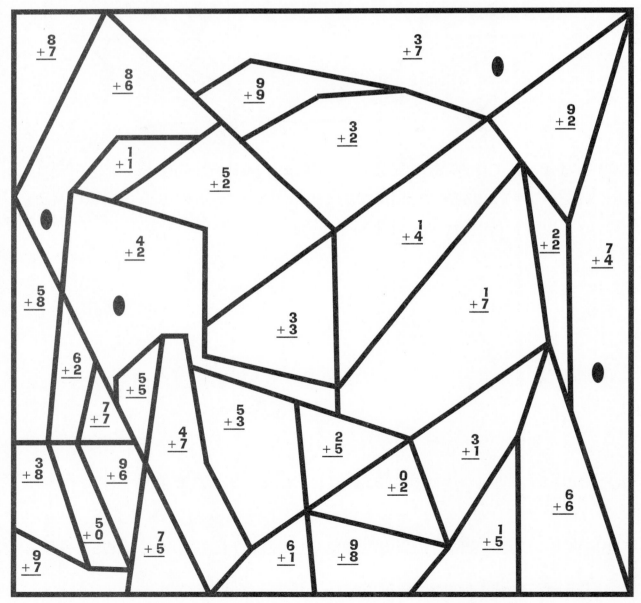

Telephone Math

Name _____ Date _____

What kind of phone never rings? _____
To find out, solve the addition problems. Then use the code
in the chart below to replace your answers with letters.
The first one has been done for you.

1 E	2 N	3 O
4 H	5 N	6 S
7 P	8 A	9 X
* H	0 T	# R

$$\begin{array}{r} 6 \\ + 2 \\ \hline 8 \end{array}$$ A

$$\begin{array}{r} 5 \\ + 1 \\ \hline \end{array}$$ ____ $$\begin{array}{r} 4 \\ + 4 \\ \hline \end{array}$$ ____ $$\begin{array}{r} 3 \\ + 6 \\ \hline \end{array}$$ ____

$$\begin{array}{r} 3 \\ + 0 \\ \hline \end{array}$$ ____ $$\begin{array}{r} 3 \\ + 4 \\ \hline \end{array}$$ ____ $$\begin{array}{r} 2 \\ + 2 \\ \hline \end{array}$$ ____

$$\begin{array}{r} 2 \\ + 1 \\ \hline \end{array}$$ ____ $$\begin{array}{r} 1 \\ + 1 \\ \hline \end{array}$$ ____ $$\begin{array}{r} 0 \\ + 1 \\ \hline \end{array}$$ ____

Write your telephone number in letters using the code above.

_____

Roger the Rooster

Name _____   Date _____

Why did Roger the Rooster decide not to get in a barnyard fight?

_____

To find out, add the numbers and shade the blocks as described below.

Shade the squares in row 1 that contain answers less than 25.
Shade the squares in row 2 that contain odd-numbered answers.
Shade the squares in row 3 that contain answers greater than 35.
Shade the squares in row 4 that contain even-numbered answers.
Shade the squares in row 5 that contain answers that end in zero.
The letters in the shaded squares spell the answer.

13 + 11 H	26 + 33 Y	16 + 31 O	10 + 12 E	64 + 24 U
20 + 15 W	71 + 12 A	25 + 21 W	51 + 10 S	22 + 16 O
22 + 10 L	14 + 14 C	20 + 10 E	25 + 31 A	21 + 3 L
42 + 30 C	13 + 43 H	54 + 15 F	21 + 61 I	61 + 33 C
10 + 30 K	20 + 30 E	16 + 32 J	71 + 23 S	70 + 20 N

Scarecrow Sam

Name _____ Date _____

Why doesn't Scarecrow Sam tell secrets when he is near Farmer Joe's bean patch? _____

To find out the answer, add the numbers. Circle the pumpkins that have sums of 14, and write the letters that appear inside those pumpkins in the boxes below.

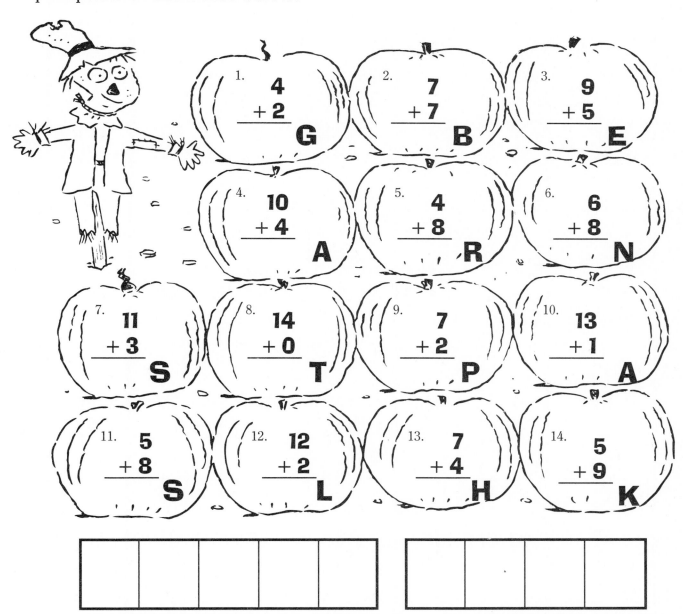

1. 4
   + 2
    G

2. 7
   + 7
    B

3. 9
   + 5
    E

4. 10
   + 4
    A

5. 4
   + 8
    R

6. 6
   + 8
    N

7. 11
   + 3
    S

8. 14
   + 0
    T

9. 7
   + 2
    P

10. 13
   + 1
    A

11. 5
   + 8
    S

12. 12
   + 2
    L

13. 7
   + 4
    H

14. 5
   + 9
    K

Number Puzzler

Name _____ Date _____

Can you spell 80 in two letters? To find out how, do the addition problems. If the answer is even, shade the square. If your answers are correct, the shaded squares will spell the answer.

12 + 13	24 + 34	22 + 21	77 + 22	35 + 43	52 + 12	40 + 52
11 + 31	30 + 39	46 + 52	15 + 12	10 + 71	63 +11	13 + 80
36 + 32	30 + 10	11 + 11	15 + 4	20 + 21	15 + 11	22 + 33
14 + 14	13 + 16	10 + 20	14 + 25	11 + 20	15 + 21	20 + 31
36 + 52	21 + 32	10 + 50	44 + 41	24 + 43	31 + 21	13 + 82

Color the Sunflower

Name _____ Date _____

Do the addition problems in the sunflower picture
below. Then use the Color Key to tell you what
color to make each answer.

Extra: Write your age on four flashcards, and then
add a 6, 7, 8, and 9 to each of the cards. Practice
the answers with a friend.

Color Key
56 = green
68 = orange
89 = yellow
97 = blue

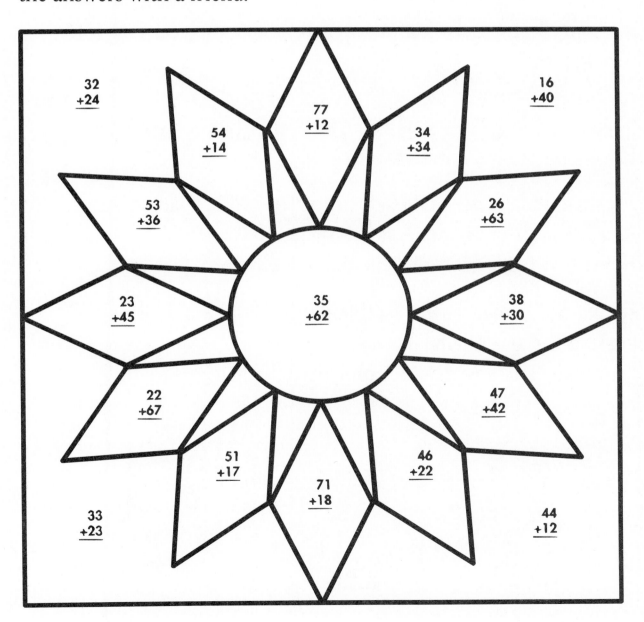

Teddy Bear Troubles

Name _____ Date _____

Solve these story problems.
Cut out the teddy bears at the bottom of the page to help you.

1. 3 teddy bears were on Jane's bed. Her sister Eva borrowed 2.
   How many teddy bears were left? _____

2. Jonathan didn't have any teddy bears. His brother Jackson gave
   him 2. His sister Shannon gave him 1.
   How many teddy bears did Jonathan have then? _____

3. Jimmy had 3 teddy bears. He thought he was too old for teddy
   bears, so he gave them all away.
   How many teddy bears did Jimmy have then? _____

28

High Flyer

Name _____ Date _____

Do the subtraction problems.

If the answer is 1 or 2,
color the shape red.
If the answer is 3 or 4,
color the shape blue.

If the answer is 5 or 6,
color the shape yellow.
If the answer is 7 or 8,
color the shape green.

If the answer is 9,
color the shape black.
Color the other shapes the
colors of your choice.

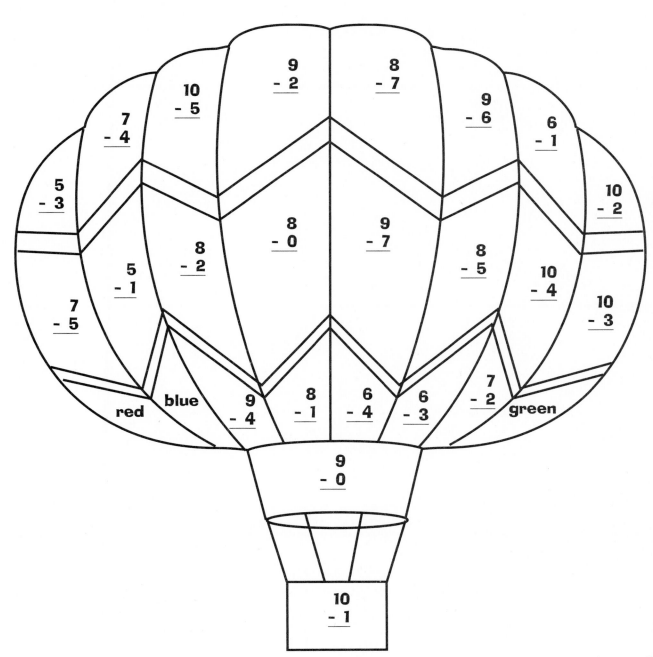

Ocean Life Subtraction

Name _____ Date _____

Use the math picture on page 31 to count and write the number in each box. Subtract the numbers.

1.  ☐
 — ☐
_____
☐

2.  ☐
 — ☐
_____
☐

3.  ☐
 — ☐
_____
☐

4. ☐
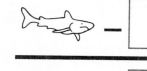 — ☐
_____
☐

5. ☐
  — ☐
_____
☐

6. ☐
 — ☐
_____
☐

7. ☐
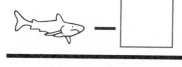 — ☐
_____
☐

8. 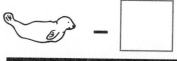 ☐
 — ☐
_____
☐

9. ☐
— ☐
_____
☐

30

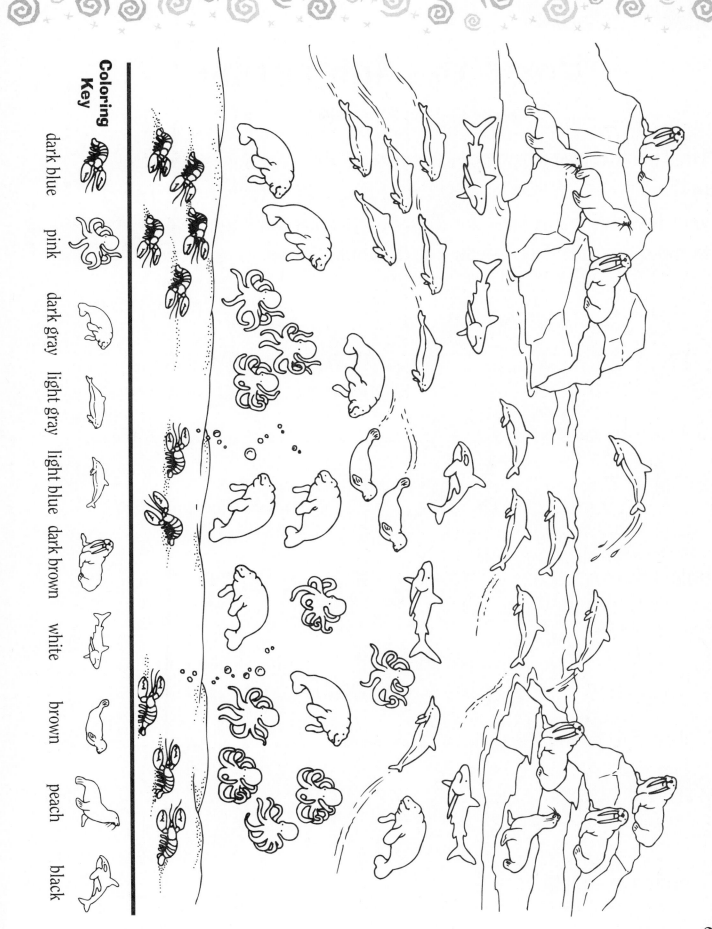

Coin-Toss Subtraction

Name _____ Date _____

Toss 3 coins. Write "H" for heads or "T" for tails in the circles below to show how the coins landed. Then finish each sentence to tell about your toss. Write a subtraction equation to show your toss, too. Write the number of heads first. We did the first one for you. Try it three times.

(H) (H) (T)   There are ___more___ heads than tails.
                        (more/fewer)

Subtraction equation: ___3 coins___ - ___2 heads___ = ___1 tail___

( ) ( ) ( )   There are _____ heads than tails.
                        (more/fewer)

Subtraction equation: _____ - _____ = _____

( ) ( ) ( )   There are _____ heads than tails.
                        (more/fewer)

Subtraction equation: _____ - _____ = _____

( ) ( ) ( )   There are _____ heads than tails.
                        (more/fewer)

Subtraction equation: _____ - _____ = _____

Baseball Puzzle

Name _____ Date _____

What animal can always be found at a baseball game?

_____

To find out, do the subtraction problems. If the answer is greater than 9, color the shapes black. If the answer is less than 10, color the shapes red.

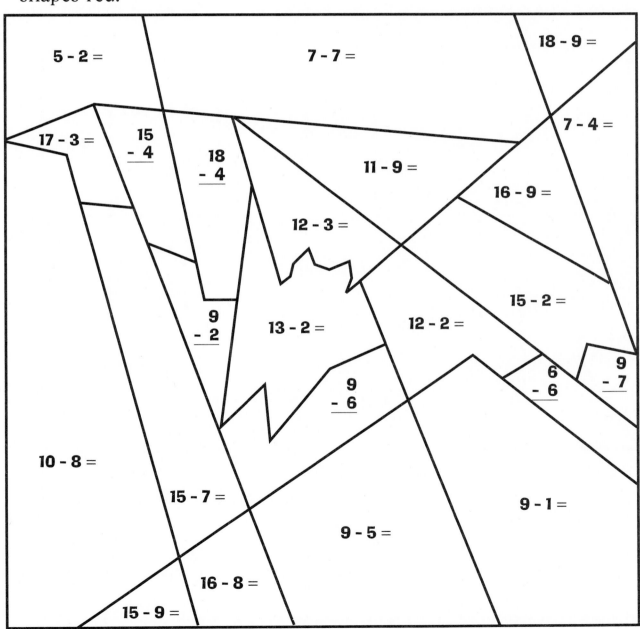

Color the Bowtie

Name _____ Date _____

Do the subtraction problems in the picture below.
Then use the Color Key to tell you what color to
make each answer.

Extra: On the back of this sheet of paper, draw a
picture of four of your friends or family members.
Give each one a bowtie!

Color Key

14 = red
26 = purple
33 = blue
47 = yellow
63 = green

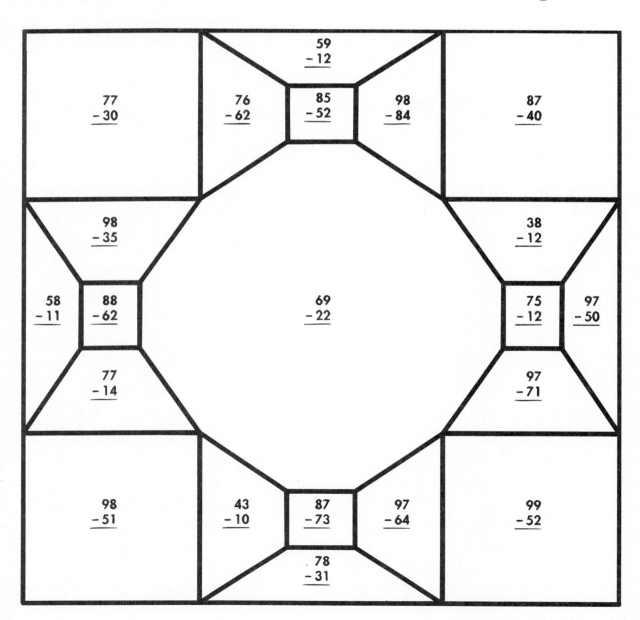

What's Your Story?

Name _____ Date _____

Look at the equation below.

$3 + 3 = 6$

Make up a story to go with the equation.

Draw a picture in the box to go with your story.

Now write about your picture on the lines below.

_____

_____

_____

_____

_____

Ice Cold Lemonade

Name _____ Date _____

Solve these story problems.
Cut out the ice cubes at
the bottom of the page
to help you.

1. Brian had a glass of lemonade.
   He added 3 ice cubes.
   The lemonade still wasn't cold.
   He added 3 more cubes.
   How many cubes did he put in the glass? _____

2. Mandy poured a glass of lemonade.
   She poured too much! The glass had 6 ice cubes.
   Mandy took out 2 cubes.
   How many cubes were left? _____

3. Travis had a glass of lemonade.
   He added 6 ice cubes.
   5 cubes melted.
   How many cubes were left? _____

36

Coin-Toss Addition

Name _____ Date _____

Toss 6 coins. Write "H" for heads or "T" for tails in the circles below to show your toss. Then write the addition equation. Write the number of "heads" first. We did the first one for you. Try it five times.

(H)(H)(H)(H)(T)(T) Equation: **4 + 2 = 6** _____

◯◯◯◯◯◯ Equation: _____

◯◯◯◯◯◯ Equation: _____

◯◯◯◯◯◯ Equation: _____

◯◯◯◯◯◯ Equation: _____

◯◯◯◯◯◯ Equation: _____

Popcorn Problems

Name _____ Date _____

Solve these story problems.
Cut out the popcorn at the bottom of the page to help you.

1. Justin went to the movies.
   After he ate some, he had 5 pieces
   of popcorn in the cup.
   He gave 2 pieces to his friend.
   How many pieces were left?

   _____

2. Ethan also went to the movies.
   His dad gave him 1 piece of popcorn.
   His mom gave him 4 pieces of popcorn.
   How many pieces did Ethan eat? _____

3. Kenya went to the school carnival.
   Her friend gave her 5 pieces of popcorn.
   Kenya ate 3 pieces.
   She fed the rest to the pet goat.
   How many pieces did she give the goat? _____

38

Butterflies on a Log

Name _____ Date _____

Solve these story problems.
Cut out the butterflies at the bottom of the page to help you.

1. A. J. saw 9 butterflies sitting on a log. He ran up to them and 7 flew away. How many butterflies were left? _____

2. Austin saw 5 butterflies flying. He saw 4 butterflies sitting on a log. How many butterflies did he see in all? _____

3. Frankie watched 9 butterflies on a log. 3 of the butterflies were monarchs. The rest were swallowtails. How many swallowtails did Frankie see? _____

39

Money Matters

Name _____ Date _____

Alex asked his little brother Billy to trade piggy banks.

Alex's bank has these coins: Billy's has these coins:

Do you think this is a fair trade? _____

Test your answer:

Add up Alex's coins: _____

Add up Billy's coins:_____

Write the totals in this Greater Than/Less Than equation:

_____ > _____

Who has more money? _____

The Truth About the Tooth Fairy

Name _____ Date _____

Look at Ali Gator's teeth.

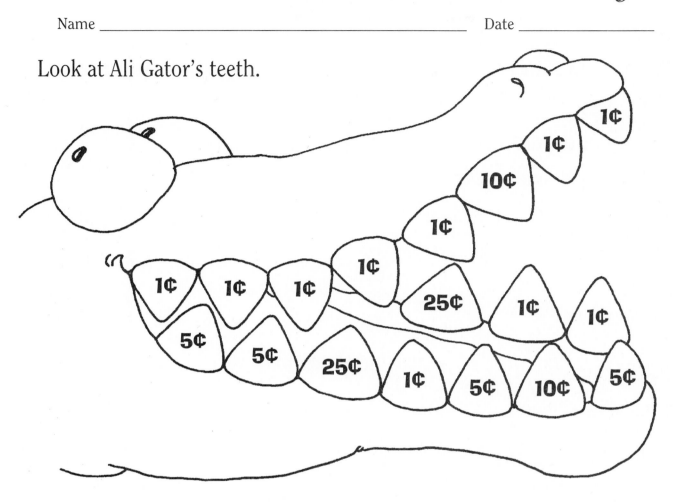

How many teeth?   How much money in all?

1. How many 1¢? [ ]   [ ] cents

2. How many 5¢? [ ]   [ ] cents

3. How many 10¢? [ ]   [ ] cents

4. How many 25¢? [ ]   [ ] cents

Measuring Up

Name _____ Date _____

People didn't always measure with rulers. Long ago, Egyptians and other peoples measured objects with body parts. Try it!

A "digit" is the width of your middle finger at the top joint where it bends. How many digits long is:

a pair of scissors? _____

a math book? _____

a crayon? _____

A "palm" is the width of your palm. How many palms long is:

a telephone book? _____

your desk? _____

a ruler? _____

A "span" is the length from the tip of your pinkie to the tip of your thumb when your hand is wide open. How many spans long is:

a broom handle? _____

a table? _____

a door? _____

Penguin Family on Parade

Name _____ Date _____

The penguin family is part of the winter parade. They need to line up from shortest to tallest. Give them a hand! Use a ruler to measure each penguin. Label each penguin with its height. Then write the name of each penguin in size order, from smallest to tallest.

Paul	**Peter**	**Patty**	**Petunia**
Height:	Height:	Height:	Height:
_____	_____	_____	_____
inches	inches	inches	inches

Size Order:

_____   _____   _____   _____

(smallest)   (tallest)

Look and Learn

Name _____ Date _____

Look at each picture. Estimate how long you think it is. Then measure each picture with a ruler. Write the actual length in inches.

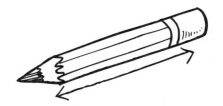

Estimate: _____ inches
Actual: _____ inches

Estimate: _____ inches
Actual: _____ inches

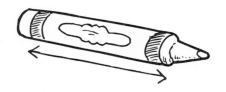

Estimate: _____ inches
Actual: _____ inches

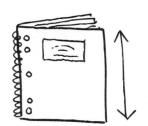

Estimate: _____ inches
Actual: _____ inches

Practice measuring other things in the room with a ruler.

Turn Up the Volume

Name _____ Date _____

How many quarts equal 1 gallon? Find out! Fill a quart container with water. Pour it into a gallon container. Keep doing it until the gallon is full. Color the correct number of quarts below. Write the numeral on the line: **1 gallon** = _____ **quarts.**

Now try it with other containers, too.

1 quart = _____ **pints**

1 pint = _____ **cups**

1 cup = _____ **tablespoons**

1 tablespoon = _____ **teaspoons**

Adding Sides

Name _____ Date _____

Use the inch side of a ruler and measure each side of each rectangle. Write the inches in the spaces below. Then add up all the sides to find the perimeter, or distance, around each rectangle.

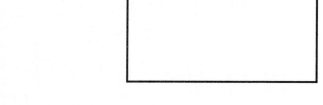

____ + ____ + ____ + ____ = _____ **inches**

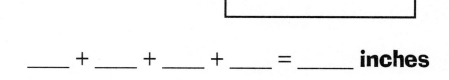

____ + ____ + ____ + ____ = _____ **inches**

____ + ____ + ____ + ____ = _____ **inches**

Measuring
length

Centimeters

Name _____ Date _____

Things can be measured using centimeters. Get a ruler that measures
in centimeters. Measure the pictures of the objects below.

book

Out Came
The Sun

_____ centimeters

book

By The River

_____ centimeters

straw

_____ centimeters

marker

_____ centimeters

5 cubes

_____ centimeters

10 cubes

_____ centimeters

shoe

_____ centimeters

hand

_____ centimeters

Five Senses

Name _____ Date _____

We learn about the world by using our 5 senses. The 5 senses are: seeing, hearing, smelling, touching, and tasting.
Look at the pictures on the left side of the graph. Think about which of your senses you use to learn about it. Draw a checkmark in the box to show the senses used. (Hint: You might use more than one.)

	See	**Hear**	**Smell**	**Touch**	**Taste**

Now graph how many senses you used for each object.

5					
4					
3					
2					
1					

Scholastic Success With Math: Grade 1

Rainbow Graph

Name _____ Date _____

Which color of the rainbow is your favorite? Color in the box for your favorite color. Have 5 classmates color the boxes to show their favorite colors, too.

Which color is liked the most? _____

Which color is liked the least? _____

Are any colors tied? _____

Which ones? _____

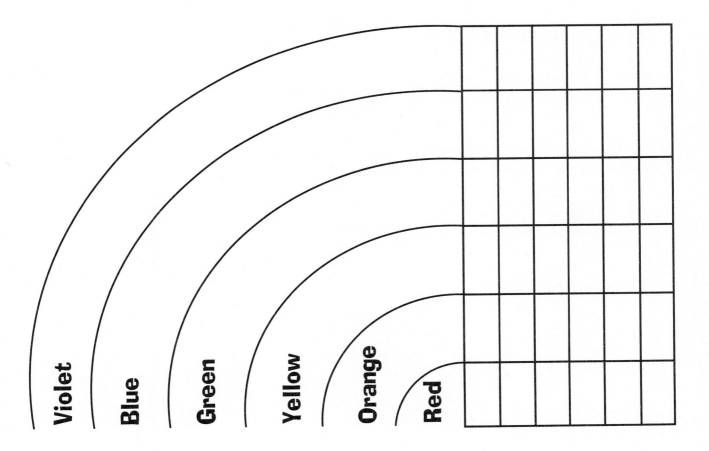

School Supplies

Name _____ Date _____

1. Find each letter and number pair on the graph.

2. Color a yellow square for each pair.

3. What picture did you make?

_____

	Across	Up			Across	Up
1.	C	4		8.	F	5
2.	C	5		9.	G	4
3.	D	4		10.	G	5
4.	D	5		11.	H	4
5.	E	4		12.	H	5
6.	E	5		13.	I	4
7.	F	4		14.	I	5

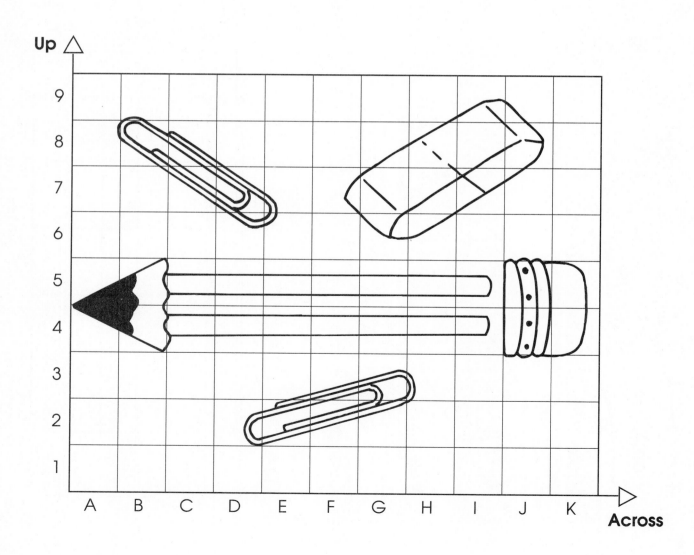

Surprises!

Name _____   Date _____

1. Find each number pair on the graph. Make a dot for each.
2. Connect the dots in the order that you make them.
3. What picture did you make?

_____

	Across	Up
1.	9	2
2.	7	4
3.	8	4
4.	6	6
5.	7	6
6.	5	8
7.	3	6
8.	4	6
9.	2	4
10.	3	4
11.	1	2

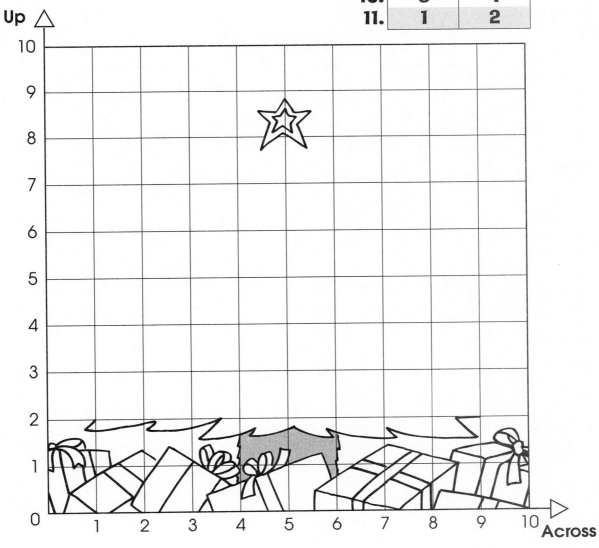

December Weather

Name _____ Date _____

In December, Mrs. Monroe's class drew the weather on a calendar.
Each kind of weather has a picture:

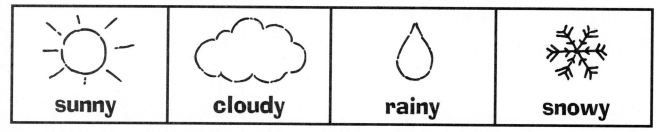

| sunny | cloudy | rainy | snowy |

Look at the calendar. Answer the questions below.

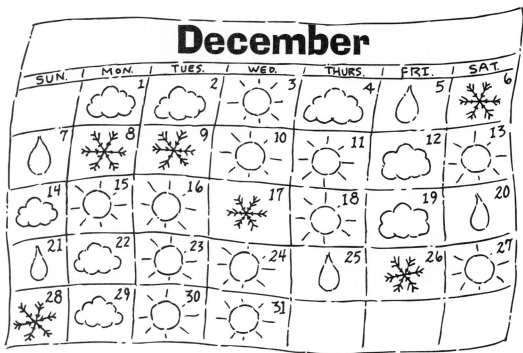

How many sunny days did they have? _____

How many cloudy days did they have? _____

How many rainy days did they have? _____

How many snowy days did they have? _____

Which kind of weather did they have the most? _____

Fun With Fractions

Name _____ Date _____

A fraction is a part of a whole.

The shapes below are split into parts, or fractions.
Color only the shapes that are split into equal parts (equal fractions).

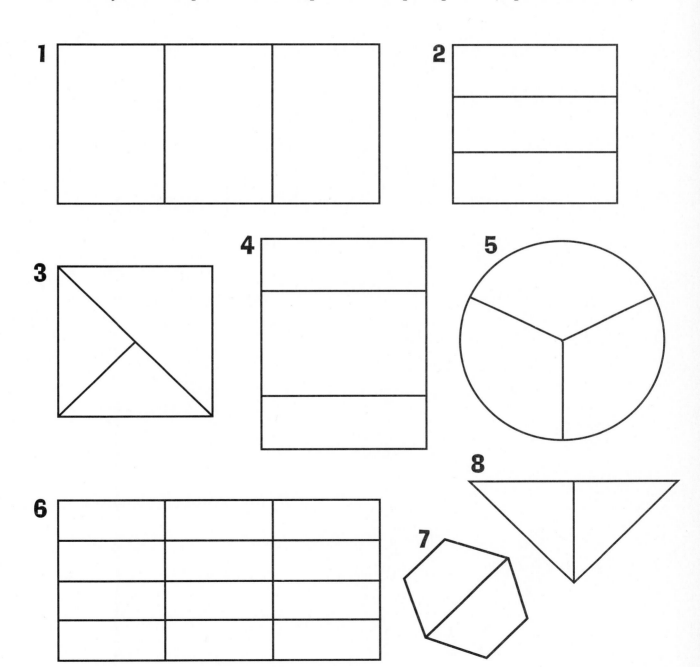

Parts to Color

Name _____   Date _____

A fraction has two numbers. The top number will tell you how many parts to color. The bottom number tells you how many parts there are.

Color 1/5 of the circle.

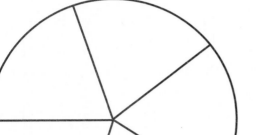

Color 4/5 of the rectangle.

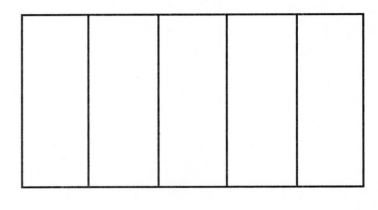

Color 3/5 of the ants.

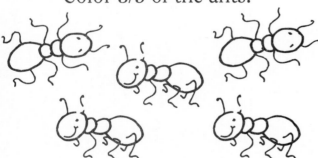

Color 2/5 of the spiders.

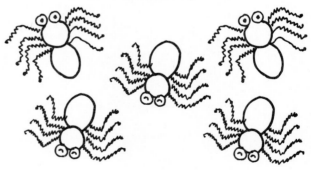

Color 0/5 of the bees.

Color 5/5 of the worms.

More Parts to Color

Name _____ Date _____

A fraction has two numbers. The top number will tell you how many parts to color. The bottom number tells you how many parts there are.

Color 1/8 of the circle.

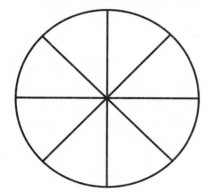

Color 6/8 of the rectangle.

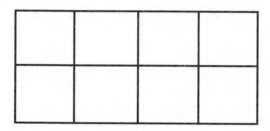

Color 4/8 of the suns.

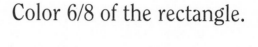

Color 8/8 of the stars.

Color 2/8 of the moons.

Color 3/8 of the planets.

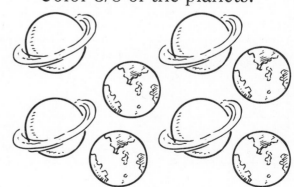

Clock Work

Name _____ Date _____

Draw the hands on the clock
so it shows 2:00.

Draw the hands on the clock
so it shows 3:00.

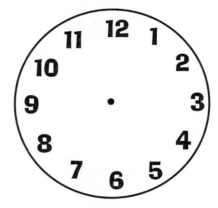

Draw the hands on the clock
so it shows 4:00.

Draw the hands on the clock
so it shows 5:00.

What do you do at 2:00 in the afternoon? Write about it on the lines below.

_____

_____

More Clock Work

Name _____ Date _____

Draw the hands on the clock so it shows 3:00.

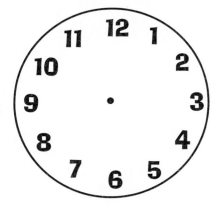

Draw the hands on the clock so it shows 6:00.

Draw the hands on the clock so it shows 9:00.

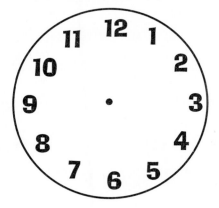

Draw the hands on the clock so it shows 12:00.

What do you do at 3:00 in the afternoon? Write about it on the lines below.

_____

_____

Even More Clock Work

Name _____ Date _____

Draw the hands on the clock so it shows 4:00.

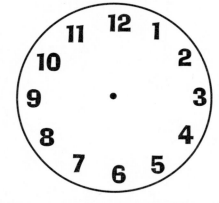

Draw the hands on the clock so it shows 4:30.

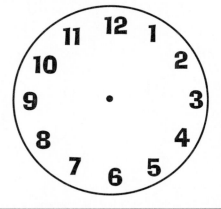

What do you do at 4:00 in the afternoon? Write about it on the lines below.

_____

_____

Draw the hands on the clock so it shows 6:00.

Draw the hands on the clock so it shows 6:30.

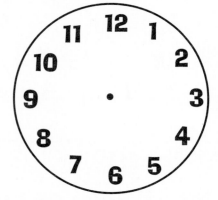

What do you do at 6:00 in the evening? Write about it on the lines below.

_____

_____

About Time

Name _____ Date _____

Why do we need to know how to tell time?
List your ideas below.

_____

_____

_____

_____

How Long Is a Minute?

Think about how much you can do in one minute.
Write your estimates in the Prediction column. Then time yourself.
Write the actual number in the Result column.

Prediction: In One Minute I Can   **Result**

Jump rope _____ times.	
Write the numbers 1 to _____ .	
Say the names of _____ animals.	

59

Answer Key

Page 5
Check children's pictures to make sure that they colored each shape the correct color: 6 = yellow; 7 = green; 8 = brown; 9 = red; 10 = green.

Page 6
Answers will vary, but check to make sure that students have supplied correct numbers for each category.

Page 7
No. Check to make sure that students have drawn lines from five different frogs to the lily pads.
Extra: 2

Page 8
1. Answers will vary.
2. Numbers will be colored in using an AB pattern of red and blue.

Page 9
Possible groups
  Balls: soccer ball, basketball, rubber ball
  Winter clothes: scarf, hat, boots
  Art supplies: paint, paintbrush, crayon

Page 10
Estimates will vary. 2, 4, 6, 8, 10, 12, 14, 16, 18, 20; 5, 10, 15, 20
Extra: No. Snowflakes would melt before you could count them.

Page 11
Greater than 10: 12 chocolate bars, 11 pennies, 15 pieces of gum, 13 candied apples
Less than 10: 7 boxes of raisins, 8 lollipops, 4 oranges, 1 cookie

Page 12
Check children's pictures to make sure that they colored each shape the correct color: one = green; two = yellow; three = red; four = purple; five = blue.

Page 13
Yield sign: triangle, 3
Caution sign: diamond, 4
Speed-limit sign: rectangle, 4
Stop sign: octagon, 8

Page 14
Left birdhouse: cube, octagon, hexagon, rectangle, square, rectangle solid
Right birdhouse: cylinder, triangle, circle, rectangle

Page 15
Color the first butterfly, the second heart, the lightbulb, and the snowflake; drawings should show the other halves.

Page 16
1. 32, 42, 52, 62, 72, 82, 92
2. 70, 60, 50, 40, 30, 20, 10
3. 67, 57, 47, 37, 27, 17, 7
4. 44, 55, 66, 77, 88, 99

Page 17
Leaf patterns will vary.

Page 18
symmetrical: heart, triangle, even V, hexagon
symmetrical twice: hexagon

Page 19

1	2	3	4	5	6	7	8	9	10
11	12	13	14	15	16	17	18	19	20
21	22	23	24	25	26	27	28	29	30
31	32	33	34	35	36	37	38	39	40
41	42	43	44	45	46	47	48	49	50
51	52	53	54	55	56	57	58	59	60
61	62	63	64	65	66	67	68	69	70
71	72	73	74	75	76	77	78	79	80
81	82	83	84	85	86	87	88	89	90
91	92	93	94	95	96	97	98	99	100

Tally marks and answers will vary.

Page 20
$4 + 4 = 8$
$5 + 5 = 10$
$6 + 6 = 12$
$7 + 7 = 14$
$8 + 8 = 16$
Extra: 6, 8, 10, 12, 14, 16
Pattern: Count by 2s, even numbers, doubling

Page 21
5 pieces
Children's patterns and equations will vary.

Page 22
$8 + 7 = 15$; $3 + 7 = 10$; $8 + 6 = 14$; $9 + 9 = 18$
$1 + 1 = 2$; $5 + 2 = 7$; $3 + 2 = 5$; $9 + 2 = 11$
$4 + 2 = 6$; $1 + 4 = 5$; $2 + 2 = 4$; $7 + 4 = 11$
$5 + 8 = 13$; $6 + 2 = 8$; $7 + 7 = 14$; $5 + 5 = 10$
$4 + 7 = 11$; $3 + 3 = 6$; $1 + 7 = 8$; $3 + 8 = 11$
$5 + 0 = 5$; $9 + 6 = 15$; $5 + 3 = 8$; $2 + 5 = 7$
$0 + 2 = 2$; $3 + 1 = 4$; $9 + 7 = 16$; $7 + 5 = 12$
$6 + 1 = 7$; $9 + 8 = 17$; $1 + 5 = 6$; $6 + 6 = 12$

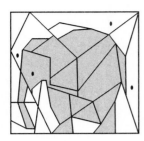

Elephant

Page 23
A SAXOPHONE
6 + 2 = 8; 5 + 1 = 6; 4 + 4 = 8
3 + 6 = 9; 3 + 0 = 3; 3 + 4 = 7
2 + 2 = 4; 2 + 1 = 3; 1 + 1 = 2
0 + 1 = 1
Phone number letters will vary.

Page 24
HE WAS A CHICKEN.
13 + 11 = 24; 26 + 33 = 59; 16 + 31 = 47
10 + 12 = 22; 64 + 24 = 88; 20 + 15 = 35
71 + 12 = 83; 25 + 21 = 46; 51 + 10 = 61
22 + 16 = 38; 22 + 10 = 32; 14 + 14 = 28
20 + 10 = 30; 25 + 31 = 56; 21 + 3 = 24
42 + 30 = 72; 13 + 43 = 56; 54 + 15 = 69
21 + 61 = 82; 61 + 33 = 94; 10 + 30 = 40
20 + 30 = 50; 16 + 32 = 48; 71 + 23 = 94
70 + 20 = 90

Page 25
BEANS TALK.
4 + 2 = 6; 7 + 7 = 14; 9 + 5 = 14
10 + 4 = 14; 4 + 8 = 12; 6 + 8 = 14
11 + 3 = 14; 14 + 0 = 14; 7 + 2 = 9
13 + 1 = 14; 5 + 8 = 13; 12 + 2 = 14
7 + 4 = 11; 5 + 9 = 14

Page 26
AT
12 + 13 = 25; 24 + 34 = 58; 22 + 21 = 43
77 + 22 = 99; 35 + 43 = 78; 52 + 12 = 64
40 + 52 = 92; 11 + 31 = 42; 30 + 39 = 69
46 + 52 = 98; 15 + 12 = 27; 10 + 71 = 81
63 + 11 = 74; 13 + 80 = 93; 36 + 32 = 68
30 + 10 = 40; 11 + 11 = 22; 15 + 4 = 19
20 + 21 = 41; 15 + 11 = 26; 22 + 33 = 55
14 + 14 = 28; 13 + 16 = 29; 10 + 20 = 30
14 + 25 = 39; 11 + 20 = 31; 15 + 21 = 36
20 + 31 = 51; 36 + 52 = 88; 21 + 32 = 53
10 + 50 = 60; 44 + 41 = 85; 24 + 43 = 67
31 + 21 = 52; 13 + 82 = 95;

Page 27
32 + 24 = 56; 16 + 40 = 56; 54 + 14 = 68
77 + 12 = 89; 34 + 34 = 68; 53 + 36 = 89
26 + 63 = 89; 23 + 45 = 68; 35 + 62 = 97
38 + 30 = 68; 22 + 67 = 89; 47 + 42 = 89
51 + 17 = 68; 71 + 18 = 89; 46 + 22 = 68
33 + 23 = 56; 44 + 12 = 56

Page 28
1. 1
2. 3
3. 0

Page 29
5 – 3 = 2; 7 – 4 = 3; 10 – 5 = 5; 9 – 2 = 7
8 – 7 = 1; 9 – 6 = 3; 6 – 1 = 5; 10 – 2 = 8
7 – 5 = 2; 5 – 1 = 4; 8 – 2 = 6; 8 – 0 = 8
9 – 7 = 2; 8 – 5 = 3; 10 – 4 = 6; 10 – 3 = 7
9 – 4 = 5; 8 – 1 = 7; 6 – 4 = 2; 6 – 3 = 3
7 – 2 = 5; 9 – 0 = 9; 10 – 1 = 9

Pages 30-31
1. 9 – 4 = 5; **2.** 8 – 1 = 7; **3.** 10 – 3 = 7
4. 8 – 6 = 2; **5.** 7 – 2 = 5; **6.** 8 – 4 = 4
7. 10 – 4 = 6; **8.** 8 – 2 = 6; **9.** 9 – 6 = 3

Page 32
Answers will vary.

Page 33
A BAT
5 – 2 = 3; 7 – 7 = 0; 18 – 9 = 9; 17 – 3 = 14; 15 – 4 = 11
18 – 4 = 14; 12 – 3 = 9; 11 – 9 = 2; 16 – 9 = 7; 7 – 4 = 3
10 – 8 = 2; 15 – 7 = 8; 9 – 2 = 7; 13 – 2 = 11; 12 – 2 = 10
15 – 2 = 13; 9 – 6 = 3; 6 – 6 = 0; 9 – 7 = 2; 15 – 9 = 6
16 – 8 = 8; 9 – 5 = 4; 9 – 1 = 8

Page 34
77 – 30 = 47; 76 – 62 = 14; 59 – 12 = 47
85 – 52 = 33; 98 – 84 = 14; 87 – 40 = 47
98 – 35 = 63; 58 – 11 = 47; 88 – 62 = 26
77 – 14 = 63; 69 – 22 = 47; 38 – 12 = 26
75 – 12 = 63; 97 – 71 = 26; 97 – 50 = 47
98 – 51 = 47; 43 – 10 = 33; 87 – 73 = 14
78 – 31 = 47; 97 – 64 = 33; 99 – 52 = 47

Page 35
Answers will vary.

Page 36
1. 6
2. 4
3. 1

Page 37
Answers will vary.

Page 38
1. 3
2. 5
3. 2

Page 39
1. 2
2. 9
3. 6

Page 40
It is not a fair trade.
Alex's coins: 5¢ + 25¢ + 10¢ = 60 ¢
Billy's coins: 10¢ + 10¢ + 10¢ + 10¢ + 10¢ +
    5¢ + 5¢ + 1¢ + 1¢ + 1¢ = 63¢
63¢ > 60¢ Billy has more money.

Page 41
1¢: 10 coins for 10¢
5¢: 4 coins for 20¢
10¢: 2 coins for 20¢
25¢: 2 coins for 50¢

Page 42
Answers will vary.

Page 43
3 1/2 inches, 2 inches, 1 1/2 inches, 3 inches
Patty, Peter, Petunia, Paul

Page 44
pencil: 2
lunchbox: 1
crayon: 2
notebook: 1

Page 45
1 gallon = **4** quarts
1 quart = **2** pints
1 pint = **2** cups
1 cup = **12** tablespoons
1 tablespoon = **3** teaspoons

Page 46
1 + 2 + 1 + 2 = **6** inches
2 + 3 + 2 + 3 = **10** inches
2 + 5 + 2 + 5 = **14** inches

Page 47
book height: 2 centimeters
book width: 3 centimeters
straw: 6 centimeters
marker: 4 centimeters
5 cubes: 4 centimeters
10 cubes: 8 centimeters
shoe: 5 centimeters
hand: 3 centimeters

Page 48
Answers will vary. The following is a likely answer.
Check children's graphs to make sure that they
correspond to the boxes checked.
chicken: see, hear, smell, touch, taste
sun: see
lemonade: see, touch, taste, smell
flowers: see, smell, touch
drums: see, hear, touch

Page 49
Answers will vary.

Page 50

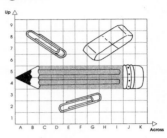

Page 51

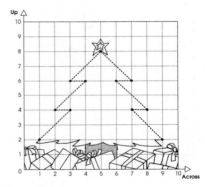

Page 52
Sunny days: 12
Cloudy days: 8
Rainy days: 5
Snowy days: 6
Most: sunny days

Page 53
Color shapes 1, 2, 5, 6, 7, and 8.

Page 54
1/5 of the circle, 4/5 of the rectangle, 3 ants,
2 spiders, 0 bees, 5 worms

Page 55
1/8 of the circle, 6/8 of the square, 4 suns, 8 stars,
2 moons, 3 planets

Page 56

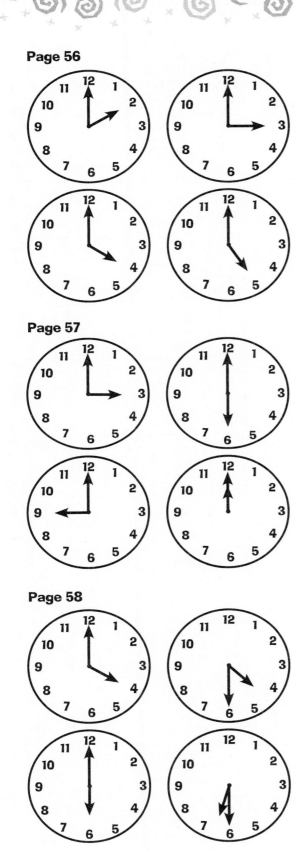

Page 57

Page 58

Page 59
Answers will vary.

Instant Skills Index